NOVICES GUIDE TO HEDGEHOG FARMING

Quillful Ventures, The Beginners Manual To Rearing, Caring And Achieving Success In Hedgehog Farming

CHERYL VALERIE

Copyright © 2024 by CHERYL VALERIE

All rights reserved. Except for brief quotations embodied in critical reviews and certain other noncommercial uses permitted by copyright law, no part of this publication may be reproduced, distributed, or transmitted in any form or by any means, Including photocopying, recording, or other electronic or mechanical methods, without the prior written permission of the publisher.

Disclaimer:

This book is a work of fiction/non-fiction, and any resemblance to actual persons, living or dead, or actual events is purely coincidental. The views expressed in this book are solely those of the author and do not necessarily reflect the views of any organization, company, or individual.

The author would like to clarify that they are not in any endorsement deal with any organization, company, or individual mentioned in this book. Any references to products, services, or entities are made for literary or informational purposes and do not constitute an endorsement or promotion.

Readers are advised to consider this work as a creative endeavor, and the author disclaims any responsibility for consequences arising from any actions or decisions based on the content of this book.

Any opinions expressed within are the author's own and should not be construed as professional advice

Contents

CHAPTER ONE..13
  COMMITTING TO HEDGEHOGS 13
    Advantages Of Farming Hedgehogs 14
    Are You A Good Fit For Hedgehog Farming?
   15
CHAPTER TWO 17
  HOW TO SET UP A HEDGEHOG FARM 17
    Selecting The Perfect Site 17
    Crucial Tools And Materials....................18
    Establishing A Cozy Ambience 19
CHAPTER THREE 21
  SPECIES AND BREEDS OF HEDGEHOG.......21
    Popular Breeds Of Hedgehogs................22
    Personality Traits And Temperaments.....23
    How To Choose The Best Hedgehog For Your Farm..25
CHAPTER FOUR..27
  SUMMARY AND HYGIENE 27
    Hedgehog Food Requirements 27
    Suggested Meals And Supplements 28

Advice On Feeding Schedules 29
CHAPTER FIVE 31
   FITNESS AND HEALTH............................31
      Common Health Concerns For Hedgehogs
     31
      Frequent Examinations And Veterinary Care
     32
      Keeping The Farm Healthy And Clean 34
CHAPTER SIX..37
   HEDGEHOG BREEDING............................37
      Comprehending Hedgehog Procreation: ..37
      Breeding Standards And Optimal
      Procedures: 38
CHAPTER SEVEN 43
   MANAGEMENT AND CONNECTIVITY...........43
      Strengthening The Link With Your
      Hedgehogs..43
      Developing Credibility With Your
      Hedgehogs..45
      Appropriate Management Methods 47
CHAPTER EIGHT..53

ETHICAL ASPECTS IN HEDGEHOG FARMING 53
   Permits And Licenses: 53
   Regulations Concerning Animal Welfare: Observance 54
   Developing A Good Rapport With Authorities: 55

CHAPTER NINE 57
  FAQS AS WELL AS TROUBLESHOOTING 57
   Taking Care Of Common Issues 57
   Housing For Hedgehogs: 58
   Concerns About Obesity: 60
   Control Of Temperature: 61

CHAPTER TEN 63
  UPCOMING IDEAS 63
   Strategic Breeding Programs 63
   Improvements To The Infrastructure 68
   Conferences And Professional Associations 70

6

ABOUT THIS BOOK

"Hedgehog Farming" is an essential resource for amateurs stepping into the fascinating field of raising hedgehogs. The introduction lays a strong basis by exploring the complexities of comprehending hedgehogs, summarizing the numerous advantages of raising hedgehogs, and providing enlightening factors to determine if this endeavor is in line with one's goals.

The second segment, "Setting Up Your Hedgehog Farm," provides a critical road map for success by highlighting the importance of picking the right site, obtaining necessary supplies and equipment, and setting up a cozy space that promotes the well-being of these adorable animals.

The third section helps readers choose the best hedgehog for their farm by providing a thorough review of common breeds, their traits, and temperaments. It does this by navigating the wide world of hedgehog species and breeds.

The fourth section discusses how important it is to carefully tend to the food requirements of hedgehogs.

To ensure the best possible health and longevity for their hedgehogs, farmers can benefit from this part, which covers everything from comprehending their dietary needs to creating a systematic feeding regimen.

The fifth section, "Health and Wellness," is an invaluable tool for addressing common health issues, stressing the importance of routine check-ups, and providing insightful information on how to keep a clean and healthy farming environment.

The book "Breeding Hedgehogs" then takes center stage, including thorough instructions on reproduction, ideal breeding procedures, and the careful attention hedgehog moms and their young need.

Building trust, employing appropriate handling skills, and providing enriching play activities for the well-being of hedgehogs are the main goals of Section 7, "Handling and Interaction," which aims to promote a harmonious connection between farmers and their spiny companions.

The eighth section delves into legal matters, emphasizing the significance of licensing, adhering to animal welfare rules, and cultivating favorable connections with authorities.

The concluding part, "FAQs and Troubleshooting," addresses common challenges, troubleshoots behavioral and health issues, and provides answers to basic FAQs so that readers are well-prepared for their adventure into hedgehog farming.

The final section, which offers a reflection on the agricultural adventure, ideas for future extension and improvement, and invaluable resources for continued learning, is where the complete character of this guide culminates.

"Hedgehog Farming" is not just a useful manual but also a lighthouse for everyone looking for a rewarding and long-lasting adventure into the fascinating world of cuddling with hedgehogs.

Overview of Hedgehog Production

Animal lovers and people looking for an unusual yet gratifying pet have come to appreciate the unique and lucrative business of hedgehog farming.

It's important to understand the basics of these entertaining animals before diving into the details of raising hedgehogs.

Small, nocturnal mammals, hedgehogs are distinguished by their distinctive spines, which act as their main line of protection. Hedgehogs are amiable, low-maintenance pets that can make people and families happy, despite their thorny exterior.

CHAPTER ONE

COMMITTING TO HEDGEHOGS

It's essential to comprehend the traits and requirements of these endearing animals before beginning a successful hedgehog farming endeavor.

As naturally occurring insectivores, hedgehogs eat mostly insects with the addition of premium commercial hedgehog food. Giving hedgehogs an appropriate habitat is essential when thinking about farming them.

To enhance their well-being, a roomy cage featuring hiding places, bedding, and a running wheel is essential.

Hedgehogs are very sensitive to temperature, thus it's critical to keep their surroundings warm and cozy for their well-being.

An important part of raising hedgehogs is reproduction.

Sows, or female hedgehogs, can give birth to four to six hoglets in each litter, on average, throughout the year.

It's imperative to give the neonates the proper care and to be well-prepared for the delivery process.

Hedgehog farming necessitates regular health check-ups, balanced food, and adequate socialization to ensure the well-being of the parents and their gorgeous young.

Advantages Of Farming Hedgehogs

Hedgehog farming has advantages that go beyond the happiness of raising these adorable animals.

Due to their relative cleanliness, hedgehogs are the perfect pet for anyone who has allergies or sensitivity to pet dander. They don't need a lot of room, therefore their small size also makes them appropriate for city living. Additionally, a hedgehog's soothing presence can lessen tension and foster a happy mood in the house.

Economically speaking, raising hedgehogs can be very profitable. The growing pet market has provided breeders with chances to contribute to the growing demand for these endearing companions. Furthermore, hedgehog farming enthusiasts may be able to generate additional revenue from hedgehog-related items including cages, bedding, and specialty food.

Are You A Good Fit For Hedgehog Farming?

It's important to determine whether hedgehog farming fits with your lifestyle and degree of commitment before getting started. Hedgehogs need devoted care, such as consistent feeding, upkeep of their habitat, and medical attention. Think about your availability and your readiness to devote time and energy to making sure these adorable animals are happy.

Hedgehog farming can be the best option for you if you want a special and fulfilling pet ownership experience and are prepared to put in the required time and money.

For those who are willing to take on the responsibility, hedgehog farming is a rewarding and worthwhile enterprise since it offers the satisfaction of watching these adorable animals flourishes under your care in addition to the possibility of financial gain.

CHAPTER TWO

HOW TO SET UP A HEDGEHOG FARM

Selecting The Perfect Site

For the health and welfare of these unusual animals, choosing the ideal site for your hedgehog farm is essential.

Hedgehogs need a steady, peaceful habitat free from loud noises and other disruptions.

Ideally, pick a space with plenty of natural light and ventilation. Make sure the temperature stays within the suggested 72 and 80 degrees Fahrenheit range.

Your hedgehogs' general comfort will be enhanced in an environment with controlled humidity levels, ideally between 40 and 60%.

Crucial Tools And Materials

Having the right materials and equipment on your hedgehog farm is essential for their well-being. Start with roomy cages that provide plenty of room to move about. To encourage hygiene, the cage should be easy to clean and escape-proof.

To satisfy their innate need to burrow, provide them with hiding places and tunnels. Invest in a top-notch heating system as well to keep the ideal temperature.

Hedgehog-friendly bedding, like fleece or recycled paper, and an appropriate exercise wheel to keep them moving are necessities.

Choose a balanced diet of commercial hedgehog food supplemented with fresh fruits, vegetables, and live insects for your hedgehog, along with plenty of water.

Make sure your hedgehogs always have access to fresh water. Check and replace their bedding regularly to keep the cages clean and free of odors.

Establishing A Cozy Ambience

Make sure your hedgehog farm is comfortable by giving it both physical and mental stimulation.

Since hedgehogs are inherently curious animals, provide toys and other enrichment materials in their cages.

Tunnels and puzzle feeders are great options to keep them occupied and interested. To promote socializing and exercise, playing outside the cage must be scheduled regularly.

To aid hedgehogs in adjusting to their new surroundings, feed and socialize with them regularly. Keep a watch on their well-being and conduct, and act quickly to address any indications of discomfort or disease. Frequent veterinary examinations are also essential to guarantee their health.

You may improve the general well-being and health of your hedgehog farm by establishing a cozy and engaging environment.

CHAPTER THREE

SPECIES AND BREEDS OF HEDGEHOG

Scientifically speaking, a wide variety of species are included in the family Erinaceidae, which includes hedgehogs. The African Pygmy Hedgehog (Atelerix albiventris) and the European Hedgehog (Erinaceus europaeus) are two of the most popular species maintained as pets or on farms. Aspiring hedgehog farmers must learn the ins and outs of each species since they each have different traits and needs.

Originating in Central and Eastern Africa, the African Pygmy Hedgehog has gained popularity because of its small size and versatility. Because of its small size (it ranges from 5 to 8 inches), it may be used in indoor and outdoor farming setups. On the other hand, the European Hedgehog, which is indigenous to

Europe, is a little bigger and has a unique look with a spiky coat that acts as insulation and armor.

Popular Breeds Of Hedgehogs

In the world of hedgehog farming, it's important to understand that breed diversity might not be as high as it is for other livestock. Individual differences in coat color and pattern are noticeable, nevertheless. The Algerian, Snowflake, Cinnamon, and Pinto hedgehog breeds are the most popular ones; they all have unique visual characteristics that enhance the aesthetic attractiveness of a hedgehog farm.

Of the breeds, the Algerian hedgehog is one of the most remarkable due to its huge white mask over its face. Conversely, snowflake hedgehogs have a distinctive coat pattern that looks like snowflakes strewn all over their spines. With their distinct cinnamon-hued quills and a combination of white and colored quills, the Pinto and cinnamon hedgehogs interestingly add to the diversity.

Personality Traits And Temperaments

Qualities and Adaptations of the Body

Comprehending the physical attributes of hedgehogs is essential for effective farming. The protective layer of spines on hedgehogs helps them regulate their body temperature in addition to acting as a deterrent to predators. Hedgehogs have short, robust legs and an acute sense of smell, which help to compensate for their relatively low vision, which extends beyond their spines. Acquainting oneself with these characteristics allows farmers to design habitats that meet their specific requirements.

Temperament Perspectives

In hedgehog farming, temperament is very important since it affects handling techniques as well as the dynamics of the farm as a whole. Hedgehogs are mostly lonesome animals that exhibit a mixture of shyness and curiosity.

On the other hand, kids can learn to tolerate social engagement if they are socialized appropriately from an early age.

It's important to recognize the unique temperaments of different breeds of hedgehogs since some may be more placid and hence good for educational settings or petting farms, while others may need more of a hands-off approach.

How To Choose The Best Hedgehog For Your Farm

Evaluating Farm's Needs

Selecting the ideal hedgehog for your farm requires carefully evaluating the traits and objectives of your farm. Take into account variables like the required amount of human interaction, the climate, and the available space. The African Pygmy Hedgehog's flexibility and manageable size make it a great option for micro indoor farms. European hedgehogs, on the other hand, might perform better in bigger, more natural outdoor environments.

Complementing Individual Preferences

The ideal hedgehog can also be chosen based on personal preferences in addition to practical factors. Certain breeds may appeal to farmers due to their distinctive appearance, while others may value handling qualities and temperament. Before deciding, it's a good idea to speak with prospective hires, observe their demeanor, and determine how well they fit the goals of the farm.

Despite being unique, hedgehog farming adheres to the same basic principles as other animal husbandry: knowing species and breeds, identifying traits and temperaments in individuals, and carefully choosing the best hedgehog for your farm based on practical and subjective factors.

Aspiring hedgehog farmers can start a fulfilling path of growing these interesting critters by carefully managing these issues.

CHAPTER FOUR

SUMMARY AND HYGIENE

Hedgehog Food Requirements

A key component of a successful hedgehog farming operation is recognizing and satisfying the dietary requirements of these prickly friends. Since hedgehogs are insectivores, high-quality protein sources should make up the majority of their diet. They consume a wide range of insects, tiny invertebrates, and occasionally even fruits in the wild. This diet must be replicated in captivity for their welfare.

Consider offering a balanced combination of commercial hedgehog food based on insects, live insects, and sporadic portions of fruits to your hedgehogs when creating their feeding schedule. They are certain to get the necessary nutrients—protein, fiber, vitamins, and minerals—thanks to this mix. Hedgehogs should not be fed dog or cat food since these foods might not include the essential nutrients that hedgehogs need to be healthy.

Suggested Meals And Supplements

Choosing the appropriate diet for your hedgehogs is essential to preserving their vigor and overall health. Choose premium commercial hedgehog food that can be found in pet stores because it is specially made to suit the dietary needs of hedgehogs. Make sure the meal you've selected has a well-balanced nutritional profile, has a sufficient amount of protein, and contains few fillers.

Another important component of hedgehog nutrition is supplementation. Your hedgehogs will get the right amount of calcium to phosphorus for strong bones if you dust them with a reliable calcium supplement while they are still alive. A multivitamin pill designed specifically for hedgehogs also helps fill up any nutritional shortages that may exist. It's crucial to avoid oversupplementing, though, as this may have negative health effects. A veterinarian's advice can assist in customizing the supplementing to your hedgehog's specific requirements.

Advice On Feeding Schedules

Creating a regular feeding routine is beneficial to your hedgehogs' general health. Since hedgehogs are nocturnal animals, it is best to feed them in the evening or at night when they are at their busiest. Hedgehogs that follow a set feeding schedule feel more secure and look forward to mealtimes.

Make sure there's fresh water in a shallow dish inside their enclosure at all times. To avoid infection, clean the water often and refill it. It's essential to keep an eye on your hedgehog's weight and modify portion amounts as necessary to avoid obesity or malnourishment. See a veterinarian right now to address any changes in feeding patterns or weight that may indicate a health issue.

It is essential to provide your hedgehogs with a diet that is both balanced and nutrient-rich for their overall well-being. You are giving your hedgehogs the best chance for a successful hedgehog farming experience by being aware of their nutritional requirements, choosing suitable meals, and maintaining a regular feeding schedule.

CHAPTER FIVE

FITNESS AND HEALTH

Common Health Concerns For Hedgehogs

Understanding the common health issues that might harm our spiny friends is essential when hedgehog husbandry.

Obesity is a common problem that is frequently brought on by overeating or insufficient exercise.

Hedgehogs are prone to putting on weight, which can cause several health issues. Make sure they eat a healthy diet and give them opportunities to move, such as an exercise wheel in their living area, to counteract this.

Another issue is skin infections, which are generally caused by filthy living circumstances. Check your hedgehogs frequently for deformities or indications of skin irritation. Quick intervention is required if it is discovered, such as modifying the bedding or cleaning the enclosure.

Keeping an atmosphere clean lowers the chance of health problems related to the skin.

Hedgehogs are also susceptible to respiratory illnesses, particularly if they are exposed to drafts or inappropriate temperatures.

It's essential to keep their dwelling area sufficiently heated and ventilated. As soon as you notice any indications of respiratory distress, get veterinarian help.

Frequent Examinations And Veterinary Care

A veterinarian's examinations regularly are essential to your hedgehog farm's residents' health. Make regular appointments to check on their general health, handle any issues, and get advice on preventive actions. Like all pets, hedgehogs gain from routine dental work, parasite control, and vaccines.

Work together closely with a licensed exotic animal veterinarian to develop a customized healthcare program for your community of hedgehogs.

 Experts will evaluate the hedgehogs for any indications of disease, dental problems, or anomalies during these examinations.

They might also talk about any apparent behavioral changes and offer advice on dietary needs.

Building a solid rapport with a reliable veterinarian will guarantee prompt intervention and preventive care, which will prolong the life and well-being of your hedgehog farm.

Keeping The Farm Healthy And Clean

The well-being of hedgehogs depends critically on a clean and healthy habitat. Keep their living area clean and sterilized regularly to stop the growth of parasites and bacteria.

Be sure to keep an eye on the bedding, the bowls for food and drink, and any accessories inside the enclosure. Keeping a regular cleaning routine reduces the chance of illness and fosters a healthy community of hedgehogs.

In addition, keep adequate ventilation to avoid respiratory problems. Enough ventilation keeps your hedgehogs comfortable and aids in temperature regulation.

To guarantee a secure and safe living environment, periodically check the enclosure and fix any possible problems.

In hedgehog farming, putting health and wellness first entails taking care of frequent health issues, making regular veterinary checkups, and keeping the farm environment clean and hygienic.

You lay the groundwork for a prosperous and happy hedgehog community by implementing these techniques into your routine for caring for hedgehogs.

CHAPTER SIX

HEDGEHOG BREEDING

Comprehending Hedgehog Procreation:

It's important to comprehend the complexities of the hedgehog reproductive cycle to breed them. Every four to six days, female hedgehogs—also referred to as sows—go into heat, which is the ideal period for successful mating. The sow releases pheromones during this time to entice possible mates. It is imperative that a breeder watch for these signals and introduce a male hedgehog, or boar, to start the mating cycle.

Hedgehogs take around 35 days to gestate, and during this time, the nutritional requirements of the pregnant sow become critical. For the health of the growing hoglets as well as the mother, it is imperative to provide a balanced diet high in protein and nutrients like folic acid. It's essential to keep an eye on the sow's weight and general health to guarantee a successful pregnancy.

Because hedgehogs are solitary creatures, it is best to keep the male and female apart after mating to prevent any possible hostility. To guarantee that the expecting mother has a cozy and peaceful place to give birth and tend to her young, it is crucial to create a cozy nesting area for her.

Breeding Standards And Optimal Procedures:

Hedgehog breeding requires careful preparation and following to recommended procedures to guarantee the animals' health and welfare. To reduce the likelihood of inherited health problems, choosing breeding pairings with strong genetic backgrounds is crucial. Before beginning the breeding process, both male and female hedgehogs must receive routine health examinations.

It is essential to provide a suitable mating environment. Allow the female hedgehog and male to socialize in a supervised and regulated environment by bringing the male inside her habitat. It's best to split up the couple after mating to protect the pregnant sow from any potential danger.

For the sake of the health of the sow and the hoglets, it is imperative to maintain an appropriate diet during the breeding process. Supplied with mealworms and fruits, premium hedgehog food helps guarantee that the mother gets all the nutrition she needs for a successful pregnancy and nursing.

Consistent observation of the gestating sow is crucial. As the due date draws near, build a comfortable nesting box with soft bedding to ensure the hoglets have a warm and safe place to be born.

A smooth and effective childbirth process is facilitated by minimizing disruptions and stress levels at this time.

Taking Care of Mother and Baby Hedgehogs:

For the hoglets to survive after birth, careful attention is essential. Although hedgehog moms are fantastic parents in general, giving them a nurturing atmosphere improves their capacity to raise children in good health. To let the mother care for and bond with her hoglets, make sure the nesting place is kept tidy, warm, and disturbance-free.

To meet her energy needs during nursing, the mother must be fed healthy food. To avoid nutritional shortages that could affect her health and the quality of her milk, include foods high in calcium in her diet. Monitor the hoglets' weight gain and overall health regularly, taking action when necessary.

As the hoglets get bigger, think about taking them away from their mother to avoid mishaps or disputes.

Give the baby hedgehogs a roomy and safe enclosure, and gently wean them off of their mother's milk.

Ensuring the hedgehog family's overall health involves keeping an eye on their development and swiftly resolving any health concerns.

A thorough knowledge of the hedgehog reproductive cycle, strict observance of breeding regulations, and committed care for both moms and their young are necessary for successful breeding.

Breeders can enjoy the pleasant pleasure of raising these endearing critters while also contributing to the health and vitality of hedgehog populations by adhering to these procedures.

CHAPTER SEVEN

MANAGEMENT AND CONNECTIVITY

Strengthening The Link With Your Hedgehogs

Hedgehog farming entails more than simply giving your prickly friends the proper climate and sustenance—it also entails developing a close relationship with them. Building trust will improve the whole farming experience and is essential to the health of your hedgehogs. Spend some time with them at first so they can become used to your voice and scent. To establish a positive association, approach their habitat calmly and speak to them in soothing tones.

Gently touch their spines, beginning at the lower back and working your way up to the head, to encourage communication. Allow them to approach you at their rate and move slowly so as not to startle them. Handling your hedgehogs regularly will help them associate your presence with good things, especially during feeding times. They will then become more open to social engagement as a result, forging a bond based on familiarity and trust.

Recognizing The Body Language Of Hedgehogs

Understanding the body language of your hedgehog is essential to effective handling. Keep a watchful eye on their quills; if they are flat and calm, they are happy. In contrast, quills that are elevated indicate dread or excitement. Keep an eye out for any discomfiting facial expressions and listen for any hissing or clicking noises. You may ensure a happy experience for both you and your hedgehogs by adjusting your handling skills based on your comprehension of these subtle signs.

Developing Credibility With Your Hedgehogs

Establishing a Cozy Ambience

Establishing a cozy and trustworthy habitat is one of the most important parts of raising hedgehogs. Start by creating a warm, haven for them to nest in that is lined with plush bedding inside of their habitat. This provides them with a secure area to retire to when they need some time alone or relaxation. It's important to maintain consistency in their environment, therefore don't make frequent adjustments to the design or décor of their habitat.

Treats are another way to foster pleasant associations during interactions. Allow them to investigate new smells and sensations at their leisure by introducing them gradually. Recall that patience is essential when establishing trust; let your hedgehogs come up to you and investigate you at their own pace. They will grow to link your presence with security and good times over time, strengthening their faith in you.

Creating a Schedule

Since hedgehogs enjoy routine, it's important to have a regular timetable for encounters. Frequent play dates, feeding schedules, and housekeeping schedules provide them with a sense of security in their surroundings. Because of this consistency and hedgehogs' increased comfort level with the daily routine, trust is fostered. Recognize that every hedgehog is different and that some may enjoy socializing in the morning, while others may be busier in the evening. Harmonious relationships are facilitated by scheduling your activities according to their innate behavioral patterns.

Appropriate Management Methods

Getting Near and Grabbing

Using the proper methods when working with hedgehogs is crucial for their welfare as well as your security. Aim to approach them slowly and deliberately; don't make abrupt moves that could frighten them. Before you try to pick them up, put your hand close to them so they can get used to your smell.

Hedgehogs can be gently scooped up from underneath by using both hands to support their bodies. To avoid any discomfort, make sure your fingers are positioned in between their quills. To give them a sense of security, hold them close to your body. If they ball up, give them time to naturally uncurl. To ensure proper handling and increase confidence, novice hedgehog farmers can benefit from practicing these procedures under the guidance of an experienced breeder or specialist.

Bathing and Nail Trimming

Responsible hedgehog farming involves routine upkeep, which includes bathing your pet sometimes and clipping your nails. To trim nails, use tiny, pet-safe clippers, being careful not to sever the nail bed. If in doubt, consult a vet or a knowledgeable hedgehog owner for advice. To bathe your hedgehog, fill a shallow basin halfway full with lukewarm water, then gradually add it, giving it time to adjust. Keep water off of their face and use a gentle shampoo that is safe for hedgehogs. When carried out carefully and consistently, these procedures improve your hedgehogs' general health.

Activities for Play and Enrichment

Establishing Intense Environments

Playtime and enrichment are essential elements of hedgehog husbandry, promoting the mental and physical health of the animals. Offer a range of hiding places, tunnels, and toys in their habitat to promote exercise and exploration. Since hedgehogs are apprehensive eaters by nature, disperse food in several areas to pique their curiosity. Provide novel textures, like cardboard or fleece blankets, to keep children interested and away from monotony.

To keep things from getting boring, occasionally rotate the toys and modify the surroundings. Providing a dynamic and engaging living environment improves their general well-being and enables them to exhibit their innate tendencies. Keep in mind that every hedgehog has distinct tastes, so you may customize their enrichment activities to fit their specific preferences by seeing how they respond to various stimuli.

Engaging Playdates

Playing interactively with your hedgehogs improves your relationship and is fun for both of you. Use a range of toys to promote exploration and mobility, such as balls and tunnels. Some hedgehogs prefer to play outside under supervision in a safe, contained area. Always keep a watchful eye on them to make sure they're safe and shielded from any dangers.

Try a variety of activities to find what engages your hedgehogs. Some could prefer a cozy play area with tunnels and barriers, while others might prefer a shallow pool for supervised swimming. Pay attention to how they react and modify playtime activities as necessary.

Playtime interactions that are consistent and constructive help your hedgehog become a contented, well-mannered pet.

Beyond just giving basic care, hedgehog farming is a fulfilling endeavor. You may establish a happy and contented atmosphere for your prickly friends by learning appropriate handling techniques, developing trust, and introducing engaging activities.

Gaining an understanding of each hedgehog's unique requirements and preferences will improve their well-being and strengthen your relationship with them, making farming a genuinely satisfying activity.

CHAPTER EIGHT

ETHICAL ASPECTS IN HEDGEHOG FARMING

Permits And Licenses:

Getting the required licenses and permits is one of the most important steps in starting a hedgehog farm. It's crucial to learn about and comprehend the particular laws controlling exotic animal farming in your area before you start this unusual agricultural adventure. Permits from environmental and agricultural agencies are usually required. Providing information regarding your farm's layout, safety precautions, and level of hedgehog care experience may be required for this.

Make sure you accurately fill out and submit all necessary documentation on time. The purpose of this procedure is to protect the environment's purity as well as the well-being of the hedgehogs. You should proceed carefully with this stage as there could be fines for not meeting licensing criteria.

Regulations Concerning Animal Welfare: Observance

Ensuring the welfare of these charming creatures is a major priority in hedgehog farming.

Following the law on animal care is not only required but also morally right. Learn about the regulations about the housing, feeding, and medical care of hedgehogs.

It is possible to perform routine inspections to make sure your farm satisfies the requirements.

To succeed in this field, invest in roomy, well-ventilated enclosures, feed your pet a healthy diet, and build a rapport with a licensed veterinarian who knows about caring for exotic animals. Exhibiting a dedication to the well-being of animals not only ensures legal compliance but also enhances the general prosperity and prestige of your hedgehog farm.

Developing A Good Rapport With Authorities:

Building a good relationship with the local government is a calculated step in the hedgehog farming industry.

Establishing a collaborative atmosphere and streamlining the regulatory process can be achieved through consistent communication and cooperation.

Request periodic inspections from the authorities, consult them for advice on compliance-related issues, and take immediate action to resolve any issues brought up.

Keeping the lines of communication open may also be helpful if industry standards or rules change.

Authorities value proactive participation, and you can help hedgehog farming maintain a strong reputation in your town by demonstrating your commitment to adhering to the law. This partnership helps the hedgehog farming business as a whole flourish sustainably while also ensuring the longevity of your farm.

CHAPTER NINE

FAQS AS WELL AS TROUBLESHOOTING

Taking Care Of Common Issues

A Beginner's Guide to Hedgehog Farming Understanding

Welcome to the world of hedgehog farming, where these cute animals offer happiness and company to a lot of people. It's normal to have doubts and worries when you set out on your adventure. To guarantee a seamless and pleasurable experience, let's address some of the most commonly asked topics and offer answers to typical problems.

Selecting the Proper Hedgehog:

How can I choose the ideal hedgehog for my farm?

In choosing a hedgehog, go for one that is curious, energetic, and not ill. Look for clear quills, a clean nose, and lively eyes. Furthermore, watch how they behave to make sure they are attentive and receptive.

Housing For Hedgehogs:

What kind of shelter is most suitable for hedgehogs?

The answer is that hedgehogs do best in large enclosures with enough airflow. To stop escapes, use a cage with vertical bars and a strong base. Provide a wheel for exercise and places to hide. Maintain a pleasant temperature range of 72–80°F (22–27°C) and keep the surroundings clean.

Nutrition and Feeding:

What kind of food is best for my hedgehog?

Since hedgehogs are omnivores, a balanced diet for them should consist of premium commercial hedgehog food along with fruits, vegetables, and insects as supplements. Steer clear of giving them dairy, sweets, or fatty foods. Fresh water should always be available.

Solving Behavioral and Health Problems

Managing Stress:

Problem: When I try to handle my hedgehog, it seems stressed and curls into a ball.

Solution: Start with brief sessions and gradually incorporate handling. Approach the hedgehog calmly and gently, giving it time to become accustomed to your scent. To link handling to good experiences, give treats during constructive interactions.

Issues with the Respiratory System:

Problem: My hedgehog is having trouble breathing and sneezing.

Solution: Environmental factors or bedding may be the cause of respiratory problems. Make sure the cage has enough ventilation, use bedding free of dust, and clean it frequently. Should symptoms continue, speak with a veterinarian.

Concerns About Obesity:

Problem: I think my hedgehog is overweight. How should I proceed?

Solution: Modify the diet by cutting back on high-calorie foods and keeping an eye on how much food is consumed. Provide more opportunities for physical activity, like a bigger wheel. See a veterinarian for a customized strategy.

Most Common Questions from Novices

Hedgehog Breeding:

I want to know if I can raise hedgehogs on my property.

The answer is that raising hedgehogs needs careful thought. Make sure you have enough room, expertise, and resources to take care of both parents and children. Maintaining the health and welfare of the hedgehog population requires responsible breeding.

Control Of Temperature:

How can I keep my hedgehogs' temperature appropriate?

To keep an eye on the temperature inside the hedgehog enclosure, use a thermometer. During the winter months, provide extra heating choices, like a ceramic heat emitter. Keep the cage away from drafty spaces.

Socialization and Interaction:

How can I strengthen my relationship with my hedgehog?

Answer: Give your hedgehog tender care, treats, and stimulating activities to help them spend meaningful time together. It takes time to establish trust, so have patience. Keep in mind that every hedgehog has a unique personality, so adjust your strategy accordingly.

CHAPTER TEN

UPCOMING IDEAS

As you commence on your hedgehog farming journey, it's vital to visualize the future ambitions of your enterprise. This requires setting clear goals, predicting probable hurdles, and strategizing for advancement. Think about your long-term goals, which can include growing your hedgehog herd, expanding your product line, or even forming alliances with nearby pet shops. By establishing your future objectives, you lay the framework for a sustainable and successful hedgehog farming business.

Strategic Breeding Programs

One major part of your future goals should include strategic breeding operations to expand the genetic variety of your hedgehog population. Selective breeding can lead to healthier and more hardy hedgehogs, minimizing the chance of genetic diseases. To share expertise and enhance your breeding methods, think about reading up on the newest methods and working with other knowledgeable hedgehog breeders.

Ecological Farming Methods

It's critical to integrate environmentally friendly techniques into your hedgehog farming business as the world's attention shifts toward sustainability. Examine possibilities for energy conservation, waste management, and sustainable habitat design. This is consistent with environmental awareness and may ultimately result in cost savings. Using sustainable farming methods improves your farm's marketability and reputation in addition to helping your hedgehog population.

Participation in the Community and Education

Consider the influence you can have on the neighborhood as well as the area around your farm. Create strategies for educating the public and engaging the community to promote responsible ownership and hedgehog care. Organizing events such as workshops, school visits, or partnerships with nearby animal shelters can enhance the perception of your farm and improve the general health of hedgehogs in your area.

Taking Stock of Your Hedgehog Farming Experience

It's essential to take stock of your hedgehog farming experience to guarantee ongoing development and flexibility. Frequent self-evaluation enables you to recognize your progress, recognize your areas of strength, and modify your approach as necessary. These are some important things to think about as you go back on your experience raising hedgehogs.

Veterinary care and health monitoring

Check the health of your hedgehog herd regularly and make sure they get veterinarian care on time. Establish a thorough health monitoring strategy to keep an eye out for any indications of disease or discomfort. Evaluate how well your present healthcare procedures are working, and be willing to try new things that will improve your hedgehogs' general health.

Efficiency of Operations

Analyze how well your farm is run daily. Think about the time and materials needed for your hedgehogs' feeding, cleaning, and supervision. Consider possible areas where operations could be streamlined or automated to maximize efficiency. This raises your productivity and raises the standard of care you give your hedgehogs.

Consumer Opinion and Contentment

To evaluate the performance of your hedgehog farming enterprise, consider the input and contentment of your customers.

Seek out feedback from customers and modify your procedures in response to their observations.

Constructive criticism offers insightful information for growth, while positive testimonials can be effective marketing tools. To systematically collect feedback, think about putting customer satisfaction surveys into place.

Future Development and Enhancements

As your hedgehog farming endeavor picks up steam, it makes sense to look into chances for growth and ongoing development. Future growth entails growing your business while continuing to provide your hedgehogs with the best care possible. Key points to keep in mind for upcoming growth and enhancements are as follows:

Improvements To The Infrastructure

Examine your current setup and note any sections that need to be upgraded to accommodate a greater number of hedgehogs. This could entail making improvements to waste management facilities, enlarging enclosures, or upgrading heating and ventilation systems. As your farm expands, infrastructure upgrades are crucial to the comfort and welfare of your hedgehogs.

Market analysis and product variety

Perform in-depth market research to find possible directions for product diversification. Examine the market for various breeds, hues, and accessories of hedgehogs. To increase the size of your market, think about working with nearby pet retailers or internet portals. In addition to generating more income, diversifying your product line allows you to serve a wider range of consumer tastes.

Employee Education and Training

As your farm grows, invest in the education and training of your employees to uphold the highest levels of care. Make sure everyone on the team is knowledgeable in caring for hedgehogs, keeping an eye on their health, and interacting with customers. Ongoing training helps your staff meet the demands of an expanding enterprise and makes your hedgehog farming enterprise more successful overall.

Resources for Lifelong Learning

Success in the ever-changing world of hedgehog farming depends on keeping up with the most recent advancements and industry best practices. The following are excellent sites for continuing education that will help you stay on the cutting edge of hedgehog husbandry:

Conferences And Professional Associations

A great way to network with seasoned breeders and keep up with industry developments is to join hedgehog breeding associations and go to pertinent conferences. These events frequently include talks, seminars, and workshops covering everything from healthcare to genetics. Engaging in such events regularly broadens your knowledge and keeps you up to date on innovative methods.

Internet Communities and Forums

Participate in online communities and forums for hedgehog breeders and hobbyists. Through platforms such as forums, social media groups, or niche websites, you can interact with a larger community, exchange experiences, and look for guidance. Engaging in these virtual forums regularly guarantees that you remain up to date with the latest developments in the hedgehog farming scene and gain access to shared knowledge.

Courses for Continuing Education

Examine courses for continuing education that are relevant to raising hedgehogs. Reputable schools provide online workshops and courses on a variety of subjects, such as business management, veterinary care, and breeding procedures. By enrolling in such classes, you may expand your knowledge and learn new abilities, which will ultimately help your hedgehog farming endeavor succeed and last.

www.ingramcontent.com/pod-product-compliance
Lightning Source LLC
Chambersburg PA
CBHW070213230526
45471CB00002B/939